9/18

Hassenfeld Library
University School of Nashville
2000 Edgehill Avenue
Nashville, TN 37212
www.usn.org

BUILDING A CAREER IN ROBOTICS

MARGAUX BAUM AND SIMONE PAYMENT

New York

Published in 2018 by The Rosen Publishing Group, Inc.
29 East 21st Street, New York, NY 10010

Copyright © 2018 by The Rosen Publishing Group, Inc.

First Edition

All rights reserved. No part of this book may be reproduced in any form without permission in writing from the publisher, except by a reviewer.

Library of Congress Cataloging-in-Publication Data

Names: Baum, Margaux, author. | Payment, Simone, author.
Title: Building a career in robotics / Margaux Baum and Simone Payment.
Description: New York : Rosen Publishing, 2018. | Series: Hands-on robotics | Audience: Grades 5–8. | Includes bibliographical references and index.
Identifiers: LCCN 2017026811| ISBN 9781499438826 (library bound) | ISBN 9781499438802 (pbk.) | ISBN 9781499438819 (6 pack)
Subjects: LCSH: Robotics—Vocational guidance—Juvenile literature.
Classification: LCC TJ211.2 .B38 2018 | DDC 629.8/92023—dc23
LC record available at https://lccn.loc.gov/2017026811

Manufactured in China

CONTENTS

INTRODUCTION 4

CHAPTER ONE
RESEARCHING ROBOTICS 7

CHAPTER TWO
COMPUTER SCIENCE 16

CHAPTER THREE
ENGINEERS AND TECHNICIANS 22

CHAPTER FOUR
ROBOTICS AT WORK 30

GLOSSARY 39
FOR MORE INFORMATION 40
FOR FURTHER READING 43
BIBLIOGRAPHY 44
INDEX 46

INTRODUCTION

Once the stuff of mere fiction or future speculation, robots have been among us for some time now, and are here to stay. The world young people will encounter in the next ten, twenty, and thirty years will probably be one where robots are not just plentiful but commonplace. Even now, robotics is a thriving technological field, and robots of one kind or another are deployed in many business sectors and areas of society.

Industrial robots have been pivotal to automobile and other heavy manufacturing efforts for decades. Ever more advanced robots utilized by surgeons, doctors, and other medical professionals for their precision in surgery and other medical support roles are coming out every year. Meanwhile, robots perform important roles in space programs, militaries, and law enforcement agencies in nations around the world.

With all the robots at work now, there will be many times more coming down the pipeline soon. Roboticists are the scientists and other specialists that help conceive of, design, and build robots and robotics system. Their jobs will be among the greatest in demand. So will the many technicians that will be necessary to manage, service, and repair a growing number of robots out there in factories, hospitals, households, and situations and environments that may be hostile to human beings, from the bottom of the sea to the outer reaches of space.

The robot revolution is infectious and has become vastly popular among young people from elementary through high school. The hobbyists of today put in long hours working on projects as part of robotics clubs, many of them going on to local, regional, national, and even international competitions, including well-known and prestigious ones like FIRST (For Inspiration and Recognition of Science and Technology) and BEST (Boosting

INTRODUCTION | 5

An engineering student at the School of Industrial Engineering of Madrid, Spain, competes in a sumo wrestling contest performed using robots during a cybertech robotics competition.

Engineering Science and Technology), where they pit their robot creations against other students'. From among their ranks will rise the next generation of roboticists, engineers, and managers.

If you are handy; have a knack for problem solving and troubleshooting; are interested in handling gear and equipment; enjoy building things and being part of a team, robotics clubs are one part of what could turn out to be an exciting career path. The other part includes getting a very good background early on in science, technology, engineering, and mathematics (STEM) subjects in school. In addition, art and design instruction won't hurt, either, since both can be applied to building robots. As students move through high school and college, they should expect to take even more advanced courses in computer science, mechanical engineering, industrial design, and more.

Robotics is a career that is poised to grow dramatically. Even now, schools are ramping up their robotics programs, new clubs are forming, and the potential jobs out there are multiplying. The students of today are lucky to live in an era in which so many of them can look forward to becoming the roboticists of tomorrow.

CHAPTER ONE

RESEARCHING ROBOTICS

At the forefront of the robotics field are the researchers pushing the envelope of cutting-edge technologies. They are the ones whose ideas and discoveries help create new robots and new applications for them, as well as improve on existing models and prototypes. Such researchers need great problem-solving abilities and must be creative, as they are often doing things that no one in the past has attempted.

TECH IMITATES LIFE

Some researchers work in the general field of robotics, and others specialize in a particular area. This means that robotics researchers might specialize in applying their knowledge of a specific science, such as biology, to robotics. A researcher with a specialization in biology might work on creating lifelike prosthetic limbs by finding ways to create robots that are modeled on biological designs. For instance, robotics researcher Sangbae Kim, who is a robot designer and an assistant professor of mechanical engineering at Massachusetts Institute of Technology (MIT), studied geckos, which are able to climb walls. Geckos have tiny hairs on their feet that act almost like suction cups. Kim used this idea to design an extremely effective robot that is able to climb walls, even those with perfectly smooth surfaces.

Researchers at universities in Switzerland and France studied how salamanders move, especially how salamanders' four feet move independently and how they switch to a different kind of movement when leaving land and entering water. This helped the researchers figure out ways to make robots with four feet move more easily over rough surfaces and led to the development of a robot known as Salamandra Robotica. Built by the Biologically

A dancing robot struts its stuff at the RoboCup Junior competition at the Institute of Robotics at Sant'Anna in Pisa, Italy, in April 2014.

RESEARCHING ROBOTICS

Inspired Robots Group, Salamandra Robotica has four feet and a body divided into nine segments. Each segment is powered by a separate microcontroller.

Some researchers radically rethink the way that robots could possibly function. For instance, robotics researcher Rodney Brooks has worked on creating robots that can figure out how to interact with their surroundings. Brooks, along with many students he has taught, is responsible for numerous advances in the field of robotics. He has built many robots, including bug-like robots and humanoid robots, and is one

Designing, engineering, and building robots can be challenging, but it is usually a collaborative endeavor, and any difficulties are often overcome via good-spirited teamwork.

of the leading researchers in the field of artificial intelligence (AI). Brooks has built small, simple, inexpensive robots based on insects that could navigate their environment. He also worked on developing a robot called Cog, which could interact with human beings, and oversaw the development of Kismet, a robot that has the ability not only to interact with people but to read and display facial expressions.

Robotics researchers often do work that is very theoretical but generally has a practical application. The possibilities inherent in the world of robotics are limitless, and robotics researchers are the people who come up with the cutting-edge robots of the future.

RESEARCHERS AT WORK

Robotics researchers are often professors at colleges and universities. An academic environment provides them with the logistical and financial support they need to conduct their research and allows them to pass on their specialized knowledge to their students in the classroom. Robotics professors might lecture to large groups of students or teach smaller classes. They will also spend time working individually with upper-level students who are working on their graduate degrees.

Robotics researchers sometimes work alone but more often form part of a larger team. They may work with robotics engineers, computer programmers, robotics technicians, or other scientists. They generally work in laboratories or at testing sites. Often able to set their own schedules, they can work nights or weekends if they choose. Robotics researchers who work at colleges or universities sometimes have summers off, at least from their teaching duties. Some choose to use this free time as a vacation, while others spend it doing further research, often

ROOMBAS AND OTHER ROBOTS: RODNEY BROOKS

Rodney Brooks is one of the world's best-known robotics researchers. He has taught and done research at Carnegie Mellon University, Stanford University, and MIT. He has also founded two robotics companies: iRobot in 1990 and Heartland Robotics in 2008. Perhaps most famously, iRobot makes the Roomba, a robotic vacuum cleaner, and it makes a number of other robots for home use. It also makes military robots such as the PackBot, a mobile robot that can scout locations, and educational robots for students and teachers. Heartland Robotics is dedicated to integrating robots into the American workforce.

Brooks is interested in finding new ways of thinking about robots and making robots that ordinary people can interact with. Obsessed with robots from the time he was a young boy, Brooks was especially interested in creating "intelligent" robots. He created a six-legged, insect-like robot named Genghis in the mid-1980s. Proposed for use in space missions, robots like Genghis had the advantage of being small, inexpensive, and able to operate autonomously. A large number of Genghis-like robots could explore a hostile environment like the surface of another planet, and the mission would not be jeopardized if a few of them were destroyed. Brooks was featured in the 1997 documentary *Fast, Cheap, and Out of Control*. Directed by Errol Morris, the film featured four people with interesting careers.

with the assistance of university students. Researchers working in academia often publish their findings in academic journals.

Some robotics researchers work for private robotics companies. At a robotics company, researchers might work on a specific project, such as finding a way to create a more efficient robotic crop harvester. Or they might work on a more general project that could have many applications, such as devising ways to improve facial recognition software that allows robots to tell the difference between individual human faces. Other robotics researchers work for the military or for defense contractors to design unmanned vehicles or for government agencies, such

This driverless car, dubbed the Pathfinder Pod and produced by UK Autodrive, is shown parked at a square in London, England, and is part of an ongoing driverless car trial underway there.

as the National Aeronautics and Space Administration (NASA), which has built and deployed robots that work on the International Space Station. The salaries that robotics researchers earn can vary greatly, depending on their experience and whether they are employed by a university or by a private company.

JOB REQUIREMENTS

Like anyone working in the field of robotics, robotics researchers will use lots of science and math in their work, so strong skills in those areas are necessary. Applying oneself to mastering those subjects in high school and college is key. It's also a good idea for students to take drafting or electronics classes and to study engineering to learn about the practical concerns that go into building a robot.

Some robotics researchers major in computer science, electrical or mechanical engineering, or even disciplines such as aeronautics in college before eventually turning their attention to robotics research. Successful robotics projects often draw on the diverse talents of a group of people who are experts in different disciplines, so having a wide base of knowledge is key.

Robotics researchers require many years of education. They generally begin their career by getting a four-year bachelor's degree, usually in a branch of science, engineering, or computing. Then they go on to earn a master's degree or a doctorate. Most robotics researchers who work as professors have at least a master's degree as well as additional postgraduate degrees.

It takes six years or more to earn a doctorate degree. While completing master's or doctorate programs, students work with professors and other students on research projects. Students can usually choose their own area of study and are supervised by a faculty member with similar interests. To complete the degree, students must write a dissertation that describes their research or give an oral presentation. With a bachelor's or master's degree, it's possible to work for robotics companies or for the government. To get a job as a robotics researcher at a university, it is usually necessary to have a doctorate.

Being a skilled communicator is especially useful for robotics researchers. Besides being a useful skill to have when teaching students, being a good communicator can help researchers express their ideas to the people they are collaborating with. Robotics researchers are often called upon to solve unique problems, so being able to think critically and come up with creative

Robotics researchers will often need to work with colleagues who are on an equal level with them on teams and also under mentors and supervisors to learn the ropes of this very technical and intensive profession.

and practical solutions is an important skill. Robotics researchers must come up with research ideas on their own, so they must be independent and able to motivate themselves.

An interest in learning is also useful for robotics researchers because most continue their education throughout their lifetimes. Beyond the basic educational requirements that are necessary to become a professional roboticist, robotics researchers should actively develop interests in other areas. Subjects such as history and biology might not seem to relate to robotics, but they can help students learn to think outside the box.

FIRST STEPS

It's never too early to start preparing for a career in robotics. Many young people build their own robots using a kit, such as the LEGO Mindstorms kit. Students should join their school's robotics team—and if their school doesn't have a robotics team, they should consider talking to school administrators and teachers about starting one. Some colleges and robotics companies sponsor high school robotics teams. There are also many activities outside of school that can be useful for preparing for a robotics career. Reading up on robots in the school library or online is a good place to start. There are many robotics summer camps in the United States that students can attend to learn valuable skills while having a great time.

CHAPTER TWO

COMPUTER SCIENCE

Robots of the past, advanced as they were for their time, were often restricted in the tasks they could do. Mostly, they were used for repetitive motions and work on factory floors. But as technology has progressed, robots have become more sophisticated. Modern robots, especially experimental prototypes, can respond to physical obstacles in their environments or human directions or actions, move independently, and even make decisions at a very basic level.

Computer scientists design and write the software that allows robots to perform functions. They might work on devising new programs and programming languages or on advanced research projects. Some computer

> One thing to prepare for in some robotics jobs is dealing with the seemingly endless lines of code needed to program even simple robotic functions and actions.

16

FROM ARTIST TO ROBOTICIST: AARON EDSINGER

Aaron Edsinger is a postdoctoral associate at MIT's Computer Science and Artificial Intelligence Laboratory. Although he started out as an artist, Edsinger got interested in building robots and the science of robotics. He went to Stanford University in California, where he earned a bachelor of science degree in computer systems and engineering. Edsinger got a master's degree—and then a PhD—in computer science from MIT. Edsinger's advisor at MIT was robotics researcher Rodney Brooks. While in college, Edsinger worked as both a research assistant and a teaching assistant.

Fascinated by the idea of creating robots that could help people, Edsinger and his collaborator Jeff Weber built Domo, a humanoid robot that can safely interact with people. Domo is equipped with a number of sensors that allow it to sense its environment, and cameras that allow it have visual perception. If handed an object, Domo can hold it up and figure out its dimensions. By interacting with people and the world around it, Domo can actually "learn" and adapt to its surroundings. Robots like Domo might someday be able to assist people with limited or impaired mobility, such as the elderly.

Edsinger co-founded two companies: Heeheehee Labs and Meka Robotics. He believes that patience is key to being a good roboticist, as developing a robot can take years. Roboticists often have to start over from scratch, and because of this, Edsinger believes it's also important for aspiring roboticists to love what they do.

scientists program individual robots or design a system to run a particular kind of robot. These computer systems might be built into the robot itself, or they might be part of a controller that operates the robot.

ROBOT PROGRAMMERS

When computer scientists are designing a program for a robot, they usually first develop a computer model. This computer model simulates both the robot and the program they are writing to control it. They continue to change and improve the program until the on-screen robot works correctly. This can take some time and involves a lot of trial and error. Computer programs

A mechanical engineering graduate student controls a robot called ESCHER at the Terrestrial Robotics Engineering and Controls Lab at Virginia Tech while preparing for a robotics contest.

rarely work correctly right from the start. It's common for computer programmers to write and rewrite computer code dozens—or even hundreds—of times. Once the computer code is working properly, it is loaded into the actual robot.

Computer scientists need to be able to think logically, have a good grasp of cause and effect, be able to break complex tasks down into smaller steps, and be able to handle multiple projects simultaneously. These professionals usually work in laboratories

FIRST IN ROBOTICS COMPETITIONS

FIRST (short for For Inspiration and Recognition of Science and Technology) is one of the premier youth organizations that promotes and runs robotics competitions. Robotics teams, which are licensed by FIRST but often affiliated with schools and their own regional and local organizations, receive kits containing parts and instructions from the organization. They then work together, under a time deadline, to build robots and compete in local competitions. Winners from local events advance to a national tournament. There they compete against students from all over the United States and from twelve other countries. Although the main competition is for students in grades nine through twelve, there are separate competitions for students in grades four through eight. Students as young as age six can compete. FIRST also runs related contests, like the FIRST LEGO League, LEGO League Jr., and FIRST Tech Challenge.

or offices, although some are able to work from home. They often collaborate on teams with robotics engineers and technicians, but must also be able to work well independently, too. To learn about the salaries computer scientists can expect to earn—and for salaries for the rest of the careers in this field, consult the Bureau of Labor Statistics' *Occupational Outlook Handbook* online (https://www.bls.gov/ooh).

COLLEGE PREP FOR COMPUTER SCIENCE

Students interested in becoming computer scientists need to be good with math and science. They should pursue their passion for computers both by both taking computer classes in high school and by working on computer projects outside of school. This can involve joining a computer club, building a web page, or even learning how to make an iPhone app.

Obviously, students will also want to be familiar with robotics and should join teams and take classes in college. To become a computer scientist, it is necessary to have at least an associate's degree in computer science or computer engineering. Most computer scientists and programmers have a bachelor's degree, and many go on to get a master's degree or a doctorate. While working toward a graduate degree, students do research projects in their area of interest. Many degree programs require that students write research papers or give presentations about their findings.

Universities with robotics programs offer various areas in which students can specialize. For example, at Stanford University's Computer Science graduate program, students can choose from a variety of computer-related specialties, including artificial intelligence. In that program, students can study how computer science is used in robotics, how machines "learn," and how machines process language.

COMPUTER SCIENCE

While early learning in junior high and high school can be intensive, it is at the college and graduate levels that the pressure will rise for many candidates hoping to build a robotics career.

Some computer science programs offer internships during the semester or help students find internships during the summer. Internships can give students a better idea of what it is like to work as a computer programmer, and working in a professional environment allows students to hone their skills and gain practical experience. Internships can also provide students with a better idea of what kind of tasks they like and where their skills lie. Many colleges help students find jobs at local companies.

CHAPTER THREE

ENGINEERS AND TECHNICIANS

Engineers who work in robotics are obviously essential in bringing robots from blueprints and drawings to real, working machines. Engineers also design, build, and program robots, as well as perform tests on them. They look at the best specifications for parts and equipment that will allow a robot to achieve its engineered purpose. Engineers solve technical glitches that arise and must have thorough and advanced training to deal with these extremely sophisticated machines. Commonly, several engineers who specialize in different aspects of robotics will team up on one project, since it is often far too much for a single individual to handle.

Robotics engineers generally focus on electrical engineering or mechanical engineering. Electrical engineers specialize in designing and developing electrical systems. Electrical engineers who specialize in robotics might design the electronics that control the robot, the robot's electronic sensors, and other electrical components. Mechanical engineers specialize in the design and development of mechanical systems. They might design the components in a robot's "hand" that allow it to pick up an object. Or, they might help develop the components that allow a robot to walk.

ENGINEERS AND TECHNICIANS | 23

A robot sprinkles cheese on a pizza as a robotics engineer looks on during a typical workday at the Institute for Artificial Intelligence at the University of Bremen, Germany, which specializes in human-scale manipulation and home-based and personal robotics.

ROBOTICS ENGINEERS ON THE JOB

Robotics engineers may focus on a particular industry, such as manufacturing. They might study the automotive industry extensively and then design robots to efficiently perform manufacturing tasks. These engineers might spend months or years designing a robotic system and then updating and revising it when it is put into use. Others may work for a company that uses robots. For example, a robotics engineer might work in an automotive plant designing the robots that work on the assembly line.

Robotics engineers are frequently called upon to work on-site. They might go to a manufacturing plant to upgrade robots or to a hospital to observe how surgeons are using robotic devices they have designed. Some robotics engineers work for a company that specializes in designing and building robots. The amount of money that robotics engineers can expect to make largely depends on what kind of company they are employed by.

FIRST RESPONDER AND ROBOT ENGINEER: ROBIN MURPHY

Robin Murphy is a robotics expert who has designed robotic devices that search for victims in disaster areas or in collapsed buildings. The robots are small enough to enter spaces where humans wouldn't fit and sturdy enough to withstand dangerous conditions. They are also smart enough to operate semi-independently of their human controllers and make some of their own decisions. A mechanical engineer with a PhD in computer science, Murphy had planned to specialize in space exploration until she realized that she could make a bigger difference in people's lives by designing rescue robots. Murphy and her robots worked at Ground Zero following the attacks on the World Trade Center in New York City on September 11, 2001, and in New Orleans after Hurricane Katrina, and they continue to help out at many other disaster sites. Murphy is a professor of computer science and engineering at Texas A&M University.

ENGINEERS AND TECHNICIANS | 25

CAREER PREP FOR ROBOTICS ENGINEERS

Robotics engineers need to be good at math and science and skilled at working with their hands, and they need to be creative thinkers who enjoy coming up with innovative solutions to problems and puzzles. They need to be disciplined and capable of motivating themselves to do a good job in a timely manner. As a part of their job, they are constantly learning new things and keeping up-to-date with the latest developments in the field.

Students interested in becoming robotics engineers should take as many algebra, geometry, and calculus classes as

Young engineering enthusiasts may be more interested in the nuts and bolts of mechanical systems than in the programming and software sides, but most robotics careers will demand wide-ranging knowledge about how robots work.

possible. They should take classes in physics and chemistry and consider taking classes in automotive mechanics, drafting, computer-aided design (CAD), and electronics. Aspiring robotics engineers that work hard can go from building robots from kits to designing actual robots as part of a design team.

In college, it's a good idea to explore a number of different kinds of engineering. Classes in computer programming are essential, as robotics engineers often need to do some programming or at least be able to work closely with computer programmers when designing or building robots. The field of robotics

Trying out different subdisciplines within robotics is a good idea for novice roboticists and is also useful for intermediate and advanced students and career-oriented professionals.

changes quickly, so engineers must always be sure to stay on top of the newest technology.

Robotics engineers need to earn a bachelor of science degree. Some colleges offer a specific degree in robotics engineering. However, most robotics engineers get a degree in another type of engineering, such as mechanical engineering or electrical engineering. While working toward that degree, they take courses in robotics. Most robotics engineers continue on to get a master's degree, and others continue on to get a doctorate. These degrees can help a robotics engineer advance his or her career. Some universities offer advanced degrees in robotics. For example, the University of Texas at Austin's Robotics Research Group offers master's and doctorate degrees in mechanical engineering with a focus on robotics.

GETTING HANDS-ON: TECHNICIANS

Robotics engineers often work alongside robotics technicians. Robotics technicians perform practical, hands-on robotics work. They build the robots and robot systems that engineers design, and repair them when they break down. They also test and maintain robots and robotic parts so they will continue to work smoothly.

Robotics technicians usually work on-site. Sometimes, the work site can be dangerous, so technicians have to be careful to observe the proper safety precautions. They may work closely with the robotics engineers who design the robots and the computer scientists who program them. They install robots in factories and set up robotic systems in laboratories. If the robots used in a manufacturing plant need to perform new tasks, robotics technicians are usually the people who make the necessary adjustments that allow this to happen. Robotics technicians may also train people who use robots. When a new robot or robotic system is installed in a factory, technicians would teach factory workers how to use and safely interact with the robot.

CAREER PREP FOR TECHNICIANS

Robotics technicians must be mechanically minded, and they should excel at using tools. Students who would like to be robotics technicians should take classes in math, automotive mechanics, CAD, drafting, electronics, and other technical

Robotics technicians can look forward to many hands-on experiences maintaining and fixing robotics systems in places where they are commonly used, including on the floors of auto plants and other industrial settings, such as this Jaguar Land Rover plant in Solihull, England.

classes. Good math and problem-solving skills are essential for this field. Most robotics technicians get an associate's degree. Others go on to get a bachelor's degree in robotics or a related field, such as computer programming or engineering. It is always a good idea to continue learning with on-the-job training or seminars.

Robotics technicians can sometimes become specialists in more than one field. For example, a robotics technician that works with automated welding systems might get training in welding as well as in robotics. In order to help program and maintain the automatic welders, the technician must know the basics of how to weld and how nonrobotic welding systems work. This additional training helps the technician in his or her current job. It also allows a technician to expand his or her knowledge base for future job opportunities.

Robotics technicians can also be certified as manufacturing technologists—people who analyze how robots and other manufacturing machines can do their jobs more efficiently within the workplace. The Society of Manufacturing Engineers offers a program that allows technicians to become a Certified Manufacturing Technologist. This involves passing a test after fulfilling school and work requirements.

CHAPTER FOUR

ROBOTICS AT WORK

Robotics will expand greatly in the coming decades, both as its own industry and within a wide array of related and even seemingly unrelated ones. Anyone who visits a hospital, stays at a hotel, orders food, or shops at a retail outlet will likely encounter robots in this new era. Military, aerospace, and industrial applications of robotics are expected to grow as well. Meanwhile, robots may become even more advanced and be able to perform certain tasks at near-human levels of proficiency if, as expected, robots are fully enhanced with artificial intelligence in the coming century.

SURGICAL AND PHARMACEUTICAL ROBOTS

Surgeons use robots to assist them in performing many types of surgery. These surgical robots can perform precise, delicate operations and have better fine-motor control than a human surgeon. Robotic surgery is also usually less invasive than traditional surgery, which allows patients to heal more quickly. One of the most popular surgical robots is the da Vinci Surgical System. When using the da Vinci system, the surgeon does not oper-

ROBOTICS AT WORK | **31**

Journalists and doctors marvel at the da Vinci Xi surgical robot at its rollout at the Gustave Roussy Institute, a leading cancer treatment center in Villejuif, France, in November 2014.

ate directly on the patient. Instead, he or she controls the robot while sitting at a computer console and watching the surgery take place on a screen in real time. If robots begin to perform far more procedures, it is likely that specialized engineering jobs that require some medical training will arise as well.

Robots are also used in some pharmacies to fill prescriptions. They locate the requested drug, dispense the correct amount of medication, and then fill and label a container. Pharmacy robots are extremely accurate and tend to cut down on errors in dispensing medication.

Robotics researchers are making many advances in improving the connection between a person's brain and a robotic limb. They hope to create advanced artificial limbs and other body parts that can function nearly as well as human limbs. There are also robots in the works that will assist nurses in performing their daily tasks, as well as robots that could aid people who live in nursing homes or need care in their own homes.

ROBOTS IN BATTLE

The military uses robots for a number of different purposes, and is constantly looking for ways to apply robotics to military applications. For instance, the military sometimes uses vehicular robots to go into dangerous areas ahead of ground troops. Some of these advanced robots are vehicles, such as the tactical unmanned ground vehicle (TUGV). Humans operate the TUGV remotely. A TUGV can enter an area before troops to check out the situation on the ground. The military also uses robotic unmanned planes. Controlled from the ground, these planes fly high over an area to perform a survey or a search. They send high-quality images back to controllers on the ground.

One experimental military robot is called BigDog. BigDog is a robot that can assist human soldiers with a number of tasks. Walking on four legs, it can carry up to 400 pounds (181 kilograms) of equipment and cargo. It can travel up to 20 miles (32 kilometers) over rough

ground and go places many vehicles cannot, such as tree-covered hillsides. BigDog can even run and faithfully follow a human leader. Scientists at Boston Dynamics are working with the Defense Advanced Research Projects Agency (DARPA), a US

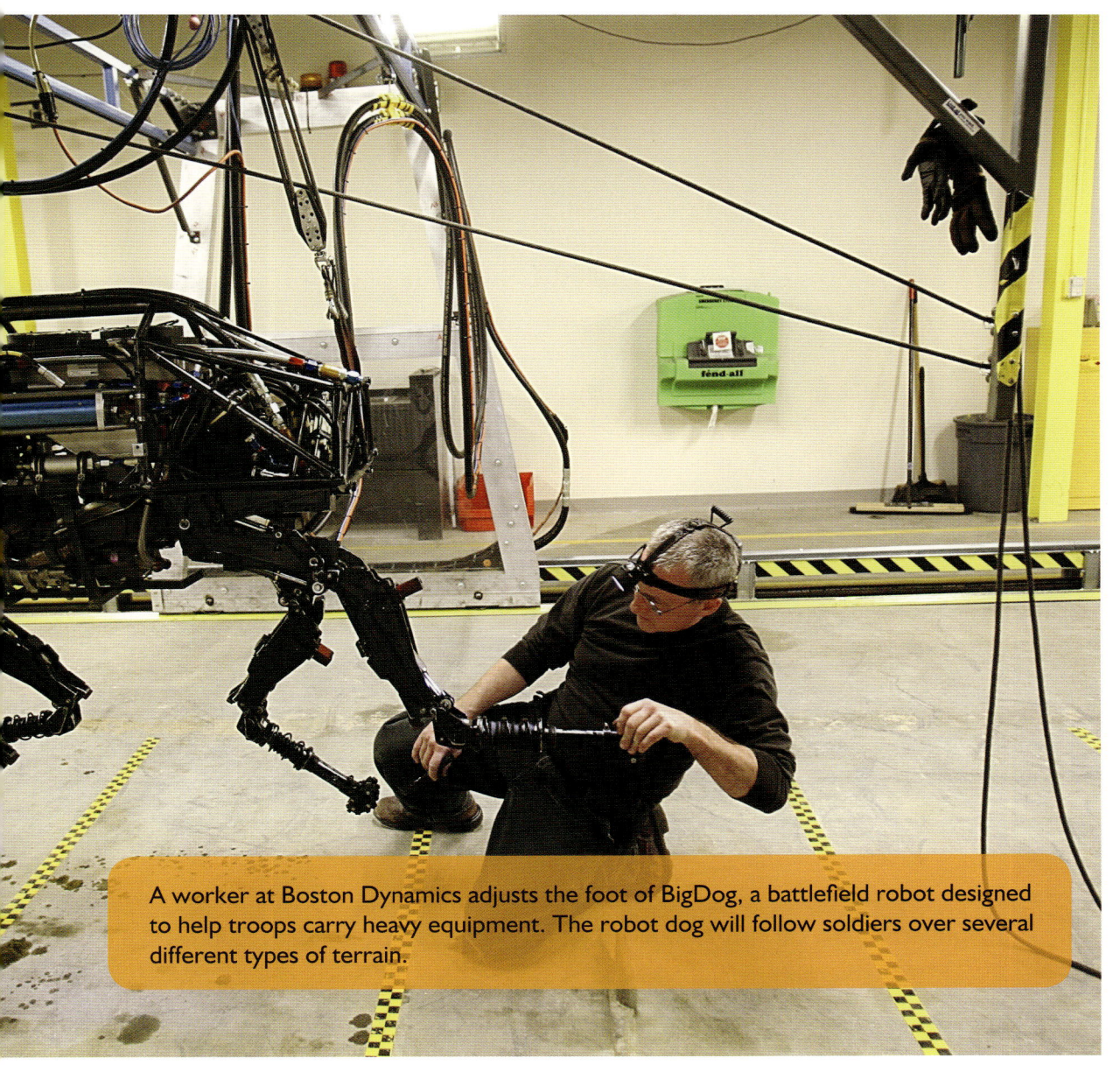

A worker at Boston Dynamics adjusts the foot of BigDog, a battlefield robot designed to help troops carry heavy equipment. The robot dog will follow soldiers over several different types of terrain.

government agency that funds technological research on behalf of the US Department of Defense. Together, Boston Dynamics and DARPA are working to make a version of the robot that can perform additional tasks, like recharging batteries. Other military robots are able to detect or blow up bombs. Some police and fire departments also use this type of robot.

AERONAUTICS AND AEROSPACE

Aeronautics and space programs around the world rely on countless robots. For instance, the International Space Station (ISS) has a robotic arm that lifts and moves equipment and cargo. The ISS also has a robot named Dextre that can work inside the station or can attach to the end of the robotic arm outside the station. Dextre can fix or replace broken parts and move small loads around the station. One major advantage of using robots on the ISS or on other space missions is that they don't need special equipment—like astronauts do—to work in the oxygen-free atmosphere of space.

Another type of space robot is a rover. Mars Exploration Rovers (MERs) have been roaming over the surface of Mars for several years. Robotics engineers on Earth control the MERs as they collect samples of the Martian soil and rocks. Onboard computers analyze the samples and send data back to Earth.

NASA is currently developing ARES, a robotic plane that may someday explore Mars. ARES is designed to parachute from a spacecraft and unfold as it falls. About 1 mile (1.6 km) from the surface, ARES will begin soaring over the Martian landscape. This will allow ARES to take photographs and collect other data.

THE ROBONAUT PROJECT

Robonaut 2 (R2) is a robot made by **NASA** and **General Motors (GM)**. The original version of Robonaut was designed to go on space missions with astronauts. **R2**, a more recent version, is considered to be one of the most advanced humanoid robots ever made. About the size of a human being, it has hands that can grip and hold objects. It is able to use the same tools that people do. **R2** also has cutting-edge sensors that allow it to tell where humans and other objects are located. **NASA** and **GM** designed R2 with safety in mind because **R2** may need to work closely with humans in dangerous situations. According to NASA, "Recently, the original upper body humanoid robot was upgraded by the addition of two climbing manipulators ('legs'), more capable processors, and new sensors."

The Robonaut 2 has a humanlike form in part because many of the spaces and equipment it has to negotiate in zero gravity are designed to human scale and for use by human space travelers.

ON THE FACTORY FLOOR

More than 150,000 robots are used in factories worldwide. A robot named Unimate was the first industrial robot. Built in 1961, it had one arm that could lift as much as 2 tons (1.8 metric tons) at a time. Unimate's job was to stack metal parts.

Today, robots used in factories are much more complex and are capable of many tasks. They perform duties such as lifting, stacking, welding, and packing. Some robots are used in factories to perform delicate jobs, like packaging cookies or candy on an assembly line. Others lift and position heavy car parts in automobile factories. Factory robots can also attach metal parts together and weld them faster and more accurately than human welders. Robots can perform the same task over and over without making mistakes, getting tired, or getting bored. They are also cleaner than humans, which can be important for types of manufacturing that require a dust-free or sterile environment. Robots also don't need to be paid a salary, and they never need to stop to eat or sleep. This can allow some factories to run for twenty-four hours a day.

They can do jobs that are too dangerous, tedious, or dirty for humans. Some robots, such as the Envirobot, manufactured by Chariot Robotics, can work with toxic chemicals and other substances. The Envirobot can quickly remove paint from steel surfaces, such as the hulls of ships, in an environmentally safe say.

ARTIFICIAL INTELLIGENCE

One of the most exciting areas for new developments in robotics is in artificial intelligence. Scientists involved in the field of AI research ways to create robots with the ability to learn, solve problems, and even reason. It's possible that someday robots with AI will learn from their experiences, just like human beings. Robots could also make their own decisions based on data they have collected.

An example of this is the Autonomous Loading System (ALS) developed at the National Robotics Engineering Center at Carnegie Mellon University Robotics Institute. The ALS is a robotic excavator that can dig up material in mines or excavation sites. It then loads trucks with excavated soil or rocks. The ALS decides how and where to dig based on what type of material it will be digging up. It then figures out the fastest and easiest path to get the excavated material into a waiting

ROBOTS AND ETHICS

From the early days of robotics, humans have considered the ethics of building and using robots. Are there types of robots humans should not design or build? Should robots be allowed to reproduce themselves? If artificial intelligence becomes advanced enough to produce very intelligent robots, will they have rights? Should robots be allowed to make decisions, such as whether to fire a weapon? Who would be responsible if a robot hurts someone? What kind of ethics should guide human beings' treatment of robots? Decisions will have to be made about these questions and many other questions we have not yet even thought of. In the future, there may be people who specialize in the ethics of robotics.

truck. The ALS scans the area to find the truck and avoid any obstacles in its path.

Another example of a robot with AI is Adam, a "robot scientist." The Adam robot was designed by a research group at Aberystwyth University in the United Kingdom and is able to conduct science experiments independently. Adam performs the first step of an experiment and then records the results. Using data from the first experiment, Adam is able to figure out what experiment to perform next. Data from the second experiment leads Adam to the next step, and so on. Adam has already made a scientific breakthrough—he found a solution to a genetics question scientists had been working to solve for years. After Adam's experiments answered this genetics question, human researchers confirmed Adam's findings.

Artificial intelligence is a rapidly developing field, and its impact on the future of robots will be substantial. In the years to come, advances in robotics are also sure to continue at a fast pace. The US Department of Labor's Bureau of Labor Statistics, a government agency that tracks job trends, expects that careers associated with robotics will continue to grow in the coming years, making robotics an excellent career choice. A career in robotics can be extremely rewarding for anyone who wants to be a part of this exciting, cutting-edge field.

GLOSSARY

dissertation A long paper written in order to earn a doctoral degree in which a student presents the results of his or her research.

doctorate An advanced degree that is earned after getting a master's degree.

drafting The act of creating formal plans or blueprints.

ethics The rules of moral conduct governing an individual or a group's actions within society.

excavate To dig something up, usually by digging a hole.

genetics The study of inherited traits.

humanoid Having human form or characteristics.

incision A cut made into the body during surgery.

internship A job performed, usually for little or no money, by students seeking to gain work experience.

locomotion The act of moving from place to place.

master's degree An advanced college degree, earned in one or more years of study after a bachelor's degree.

pharmacy A place where prescription medicine is prepared and distributed.

postgraduate Schooling completed after completion of college.

prescription A written direction or order for the preparation and use of a medicine.

prosthesis An artificial device or part to replace missing or injured limbs or other parts of a human body.

tournament A contest or series of contests played to determine an ultimate champion.

trait A quality or characteristic of something.

FOR MORE INFORMATION

Boosting Engineering, Science, and Technology (BEST) Robotics, Inc.
PO Box 1024
(334) 844-5759
Website: http://best.eng.auburn.edu
BEST is a nonprofit, volunteer-based organization whose mission is to inspire students to pursue careers in engineering, science, and technology through participation in a sports-like, science- and engineering-based robotics competition.

Canadian Space Agency
John H. Chapman Space Centre
6767 Route de l'Aéroport
Saint-Hubert, QC J3Y 8Y9
Canada
(450) 926-4800
Website: http://www.asc-csa.gc.ca/eng/default.asp
Facebook: @CanadianSpaceAgency
Instagram: @canadianspaceagency
The Canadian Space Agency is responsible for Canada's space missions, including robotics work on the International Space Station.

FIRST Robotics Canada
PO Box 518
Pickering Main
Pickering, ON L1V 2R7
Canada
(416) 396-5907
Website: http://www.firstroboticscanada.org

FOR MORE INFORMATION | 41

FIRST Robotics Canada sponsors robotics contests for youth throughout Canada.

FIRST Robotics Competitions
200 Bedford Street
Manchester, NH 03101
(800) 871-8326
Website: https://www.firstinspires.org
Twitter: @FIRSTweets
FIRST Robotics Competitions are among the most well known of their kind nationwide and internationally, and their website provides information about past challenges, upcoming contests, and scholarship opportunities.

IEEE Robotics and Automation Society
445 Hoes Lane
Piscataway, NJ 08854
(732) 562-3906
Website: http://www.ieee-ras.org
Twitter: @ieeeras
A national society, the Robotics and Automation Society is part of the Institute of Electrical and Electronic Engineers (IEEE) and offers professional and student chapters that focus on advancing innovation, education, and research dealing with robotics and automation.

Robotics Alliance Project
National Aeronautics and Space Administration (NASA)
NASA Headquarters
300 E Street SW
Washington, DC 20546-0001
(202) 358-0001
Website: http://robotics.nasa.gov

The Robotics Alliance Project is a branch of NASA that focuses on getting students interested in and prepared for robotics careers in the space industry.

Robotics Institute
Carnegie Mellon University
5000 Forbes Avenue
Pittsburgh, PA 15213-3890
(412) 268-3818
Website: http://www.ri.cmu.edu
The Robotics Institute at Carnegie Mellon University has college programs for students interested in pursuing a career in robotics. It also has summer camp robotics programs for younger students.

Technology Student Association (TSA)
1914 Association Drive
Reston, VA 20191
(703) 860-9000
Website: http://www.tsaweb.org
The TSA is a national nonprofit educational organization of students who are interested in STEM, with student-led chapters spanning forty-nine states.

WEBSITES

Due to the changing nature of internet links, Rosen Publishing has developed an online list of websites related to the subject of this book. This site is updated regularly. Please use this link to access the list:
http://www.rosenlinks.com/HOR/Career

FOR FURTHER READING

Asimov, Isaac, and Richard Hantula. *Science Fiction: Vision of Tomorrow?* Milwaukee, WI: Gareth Stevens Publishing, 2005.

Chaffee, Joel, and Margaux Baum. *Engineering and Building Robots for Competition* (Hands-on Robotics). New York, NY: Rosen Publishing, 2018.

Cook, David. *Robot Building for Beginners*. 2nd ed. New York, NY: Apress, 2010.

Freedman, Jeri, and Margaux Baum. *The History of Robots and Robotics* (Hands-on Robotics). New York, NY: Rosen Publishing, 2018.

Greek, Joe. *Artificial Intelligence* (Digital and Information Literacy). New York, NY: Rosen Publishing, 2018.

Hustad, Douglas. *Discover Robotics*. Minneapolis, MN: Lerner Publications, 2017.

La Bella, Laura. *The Future of Robotics* (Hands-on Robotics). New York, NY: Rosen Publishing, 2018.

Mara, Wil. *Robotics Engineers* (Cool STEAM Careers/21st Century Skills Library). Ann Arbor, MI: Cherry Lake Publishing, 2015.

Peppas, Lynn. *Robotics*. New York, NY: Crabtree Publishing, 2015.

Piddock, Charles. *Future Tech: From Personal Robots to Motorized Monocycle*. Washington, DC: National Geographic Society, 2009.

Ryan, Peter K. *Powering Up a Career in Robotics* (Preparing for Tomorrow's Careers). New York, NY: Rosen Publishing, 2015.

Spilsbury, Louise, and Richard Spilsbury. *Robotics*. New York, NY: Gareth Stevens Publishing, 2017.

BIBLIOGRAPHY

Boston Dynamics. "Big Dog: The Most Advanced Rough-Terrain Robot on Earth." Retrieved March 10, 2009. http://www.bostondynamics.com/robot_bigdog.html.

Careers in Focus: Engineering. New York, NY: Ferguson/Infobase Publishing, 2007.

Dimberu, Peniel M. "Adam the Robot Scientist Makes Its First Discovery." Singularity Hub. Retrieved March 16, 2010. http://singularityhub.com/2010/03/16/adam-the-robot-scientist-makes-its-first-discovery.

Discovering Careers for Your Future: Space Exploration. New York, NY: Ferguson/Infobase Publishing, 2008.

Cohn, Jessica. *Top Careers in Two Years: Manufacturing and Transportation.* New York, NY: Ferguson Publishing, 2008.

Hyland, Tony. *Robots at Work and Play.* North Mankato, MN: Smart Apple Media, 2008.

Hyland, Tony. *Space Robots.* North Mankato, MN: Smart Apple Media, 2008.

Kupperberg, Paul. *Careers in Robotics.* New York, NY: Rosen Publishing Group, 2007.

Leggett, Hadley. "Deep-Sea Robot Roves the Unexplored Ocean Depths." *Wired*, Science. September 11, 2009. http://www.wired.com/wiredscience/2009/09/benthicrover.

Mataric, Maja M. *The Robotics Primer.* Cambridge, MA: MIT Press, 2007.

NASA. "R2." Retrieved February 25, 2017. http://robonaut.jsc.nasa.gov.

National Robotics Engineering Center. "Autonomous Loading System (ALS)." Retrieved March 20, 2017. http://www.nrec.ri.cmu.edu/projects/als..

O-NET OnLine. "Summary Report for: 17-3024.01—Robotics Technicians." Retrieved February 3, 2009. http://online.onetcenter.org/link/summary/17-3024.01.

Piddock, Charles. *Future Tech: From Personal Robots to Motorized Monocycle*. Washington, DC: National Geographic Society, 2009.

Rothman, Wilson. "Rescuer by Remote: Need Help? Send in the Robot." *Time*, June 8, 2004. http://www.time.com/time/2004/innovators/200406/murphy.html.

Schneider, David. "Robin Murphy: Roboticist to the Rescue." IEEE Spectrum: Inside Technology. February 2009. http://spectrum.ieee.org/robotics/artificial-intelligence/robin-murphy-roboticist-to-the-rescue.

Steffoff, Rebecca. *Robots*. New York, NY: Marshall Cavendish, 2008.

Tech Museum of Innovation. "Machines and Man: Ethics and Robotics in the 21st Century." Retrieved February 3, 2009. http://www.thetech.org/robotics/ethics/index.html.

Trafton, Anne. "From Nature, Robots: Mechanical Engineer Sang-bae Kim Looks to Animals to Inspire His Robot Designs." MIT News, September 25, 2009. http://web.mit.edu/newsoffice/2009/stickybot-092509.html.

INDEX

A

Adam, 38
aerospace and robotics, 34
ARES, 34
artificial intelligence (AI), 20, 30, 36–38
Autonomous Loading System (ALS), 37–38

B

BigDog, 32–33
Biologically Inspired Robotics Group, 8–9
biology and robotics, 7–9
Boosting Engineering Science and Technology (BEST) 4–5
Boston Dynamics, 33–34
Brooks, Rodney, 9–10, 11

C

Cog, 10
computer science and robotics, 16, 18–20
  educational requirements, 20–21
  internships, 21

D

da Vinci Surgical System, 30–31
Defense Advanced Research Projects Agency (DARPA), 33–34
Dextre, 34
Domo, 17

E

Edsinger, Aaron, 17
Envirobot, 36

F

Fast, Cheap, and Out of Control, 11
For Inspiration and Recognition of Science and Technology (FIRST), 4, 19

G

Genghis, 11

H

Heartland Robotics, 11

I

industry/factories and robotics, 36
International Space Station, 34
iRobot, 11

K

Kim, Sangbae, 7–8
Kismet, 10

M

Mars Exploration Rovers (MERs), 34
military and robotics, 32–34
Murphy, Robin, 24

INDEX

N
National Aeronautics and Space Administration (NASA), 13, 35

O
Occupational Outlook Handbook (Bureau of Labor Statistics), 20

P
PackBot, 11

R
Robonaut 2 (R2), 35
roboticist, definition of, 4
robotics careers
  educational requirements, 13–14, 25–27
  engineers, 22–27
  getting started, 15
  increasing demand, 4–6, 30, 38
  researchers, 10, 12–13
  salary, 13, 24
  skills needed, 7, 14–15
  technicians, 27–29
robots
  ethics and, 37
  evolution of, 16
  used in various economic sectors, 4
Roomba, 11

S
Salamandra Robotica, 8–9
science, technology, engineering, and mathematics (STEM), 5
surgery/medicine and robotics, 30–32

T
tactical unmanned ground vehicle (TUGV), 32

U
Unimate, 36

W
Weber, Jeff, 17

ABOUT THE AUTHOR

Margaux Baum is an author and editor from New York with many published credits specializing in technology and careers.

Simone Payment has a degree in psychology from Cornell University and a master's degree in elementary education from Wheelock College. She is the author of twenty-four books for young adults. Her book *Inside Special Operations: Navy SEALs* (also from Rosen Publishing) won a 2004 Quick Picks for Reluctant Young Readers award from the American Library Association and is on the Nonfiction Honor List of Voice of Youth Advocates.

PHOTO CREDITS

Cover Peter Cade/Iconica/Getty Images; p. 5 Pablo Blazquez Dominguez/Getty Images; p. 8 Laura Lezza/Getty Images; p. 9 nd3000/Shutterstock.com; pp. 11, 17, 19, 24, 35, 37 (background) Sylverarts Vectors/Shutterstock.com; p. 12 Anadolu Agency/Getty Images; p. 14 zoranm/E+/Getty Images; p. 16 kikujungboy/Shutterstock.com; p. 18 Chip Somodevilla/Getty Images; p. 21 izusek/E+/Getty Images; p. 23 Ingo Wagner/DPA/Getty Images; p. 25 Hero Images/Getty Images; p. 26 © iStockphoto.com/DGLimages; p. 28 Leon Neal/Getty Images; p. 31 Francois Guillot/AFP/Getty Images; pp. 32–33 Boston Globe/Getty Images; p. 35 NASA.

Design: Michael Moy; Photo Research: Karen Huang

USN

Hassenfeld Library
University School of Nashville
2000 Edgehill Avenue
Nashville, TN 37212
www.usn.org